I0796308

Early
NATURAL DISASTERS
Encyclopedias

HURRICANES

by Carla Mooney

Early Encyclopedias

An Imprint of Abdo Reference
abdobooks.com

abdobooks.com

Published by Abdo Reference, a division of ABDO, PO Box 398166, Minneapolis, Minnesota 55439.

Printed in China.
102024
012025

Editor: Christa Kelly
Series Designers: Candice Keimig, Joshua Olson
Production Designer: Joshua Olson

Library of Congress Control Number: 2024938363

Publisher's Cataloging-in-Publication Data

Names: Mooney, Carla, author.
Title: Hurricanes / by Carla Mooney
Description: Minneapolis, Minnesota: Abdo Reference, 2025 | Series: Early natural disasters encyclopedias | Includes online resources and index.
Identifiers: ISBN 9781098296049 (lib. bdg.) | ISBN 9798384917045 (eBook)
Subjects: LCSH: Hurricanes--Juvenile literature. | Natural disasters--Juvenile literature. | Weather--Juvenile literature. | Severe storms--Juvenile literature. | Meteorology--Juvenile literature. | Earth science--Juvenile literature. | Encyclopedias and dictionaries--Juvenile literature.
Classification: DDC 551.552--dc23

CONTENTS

INTRODUCTION

Hurricanes combine wind, rain, and tornadoes into one dangerous storm.

What Are Hurricanes?

Hurricanes are some of the most powerful storms on Earth. They can stretch across hundreds of miles. Hurricanes form over oceans. These storms get energy from an ocean's warm waters.

Deadly Storms

Hurricanes are dangerous. They threaten property. They put lives in danger. Hurricanes have fast winds. These winds can rip roofs off houses. They can destroy buildings. Hurricanes bring heavy rain. They also bring high ocean levels called storm surge. This can cause flooding.

Hurricanes can travel hundreds of miles inland.

Parts of a Hurricane

A hurricane's winds rotate around a center called an eye. The weather in the eye is mostly calm. An eyewall surrounds the eye. The hurricane's strongest winds are in the eyewall. Reaching out from the eyewall are rainbands. Rainbands are curved arms of rain and wind.

Hurricane eyes can be shaped like circles, triangles, squares, or pentagons.

Inside a Hurricane

Hurricanes have an eye, an eyewall, and rainbands.

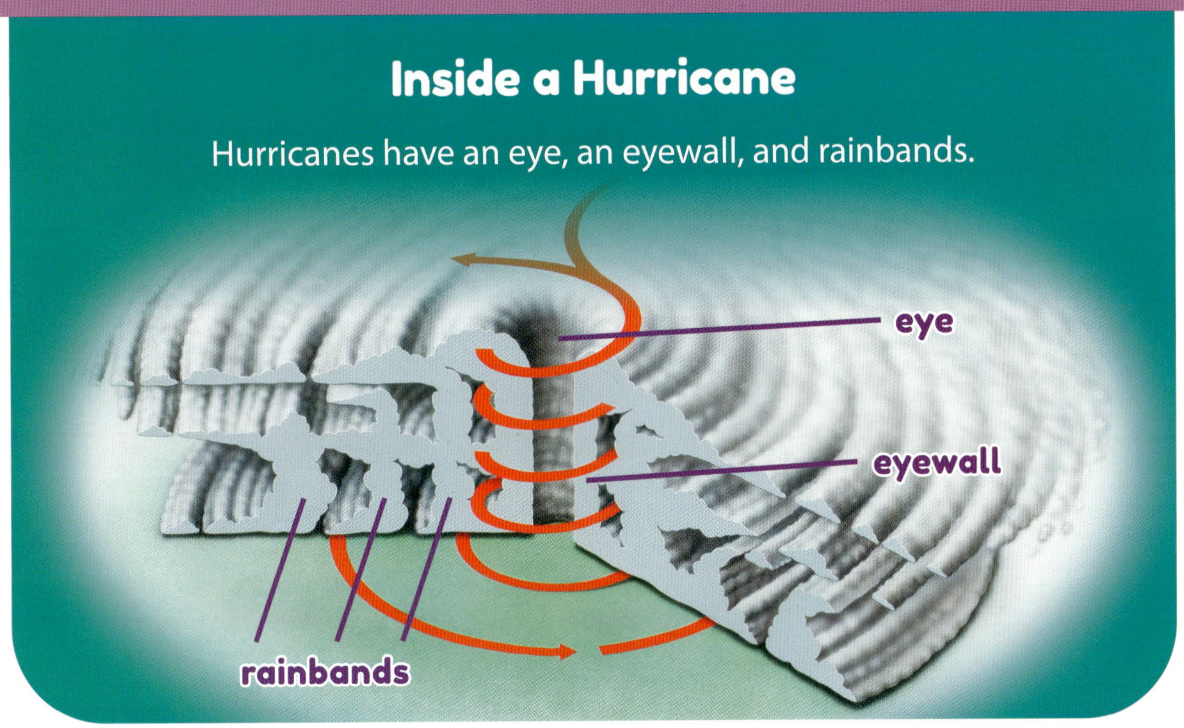

Hurricanes around the World

Hurricanes have different names around the world. They are called hurricanes in the Atlantic Ocean and the central and northeastern Pacific Ocean. They are known as typhoons in the northwestern Pacific. They are called cyclones in the Indian Ocean and southern Pacific Ocean.

FUN FACT!

The Taino are Indigenous people of the Caribbean. The word *hurricane* comes from a Taino word that means "evil spirit of the wind."

Ocean storms begin when ocean water evaporates and forms clouds.

The Storm Begins

A hurricane starts as a storm over warm ocean water. Humid air near the ocean's surface rises. The water vapor in the air cools as it rises. The cooling air forms storm clouds. The water vapor in the clouds condenses. It becomes rain droplets.

Heating and Cooling

The rising air releases heat as it cools. The heat warms the air in the storm. Air near the ocean's surface flows underneath the rising air to replace it. The cycle continues. The storm grows.

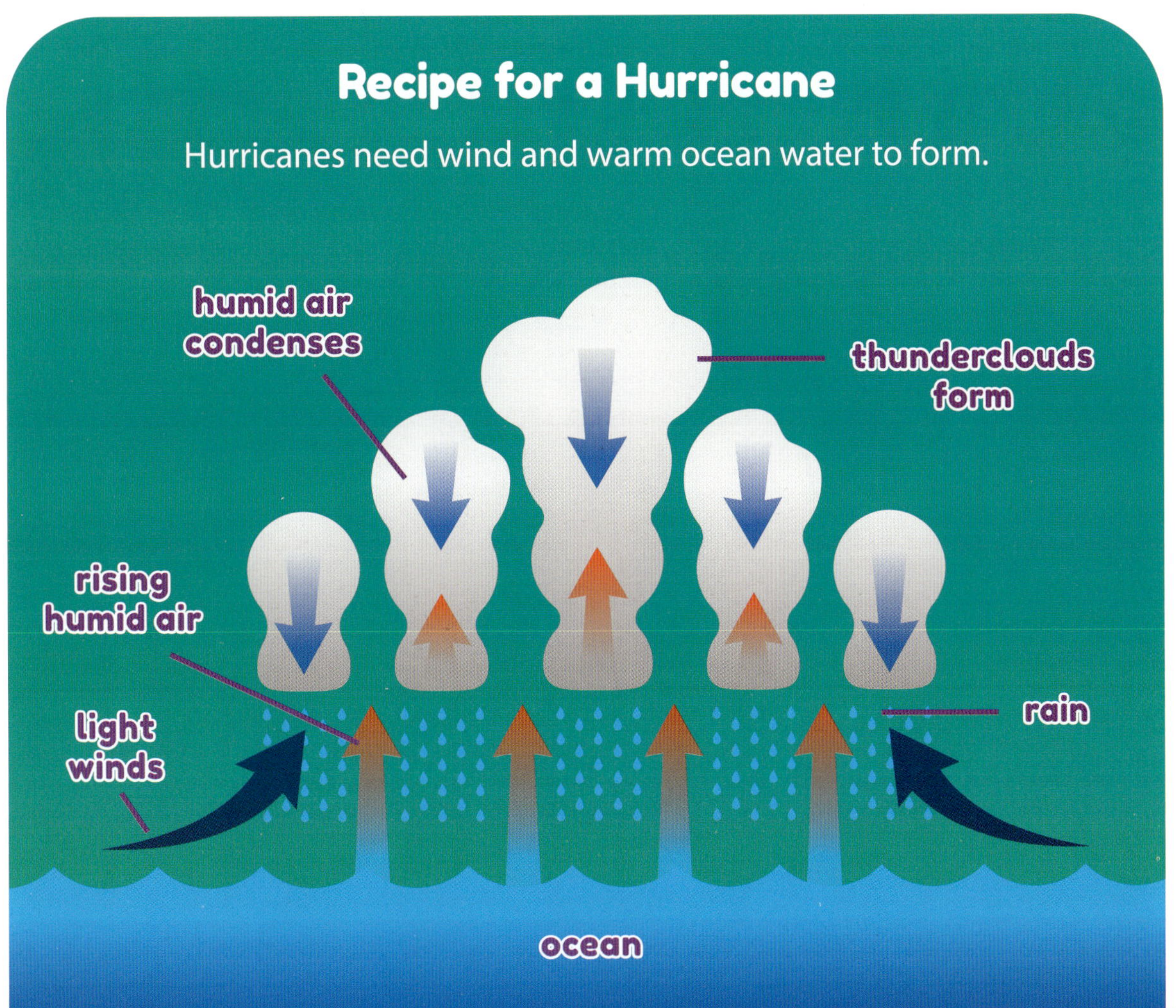

A Tropical Disturbance

A group of clouds and thunderstorms over warm ocean waters is called a tropical disturbance. Most tropical disturbances weaken and fade away. But some grow.

Ocean storms can form over cool ocean waters, but those storms are not considered tropical disturbances.

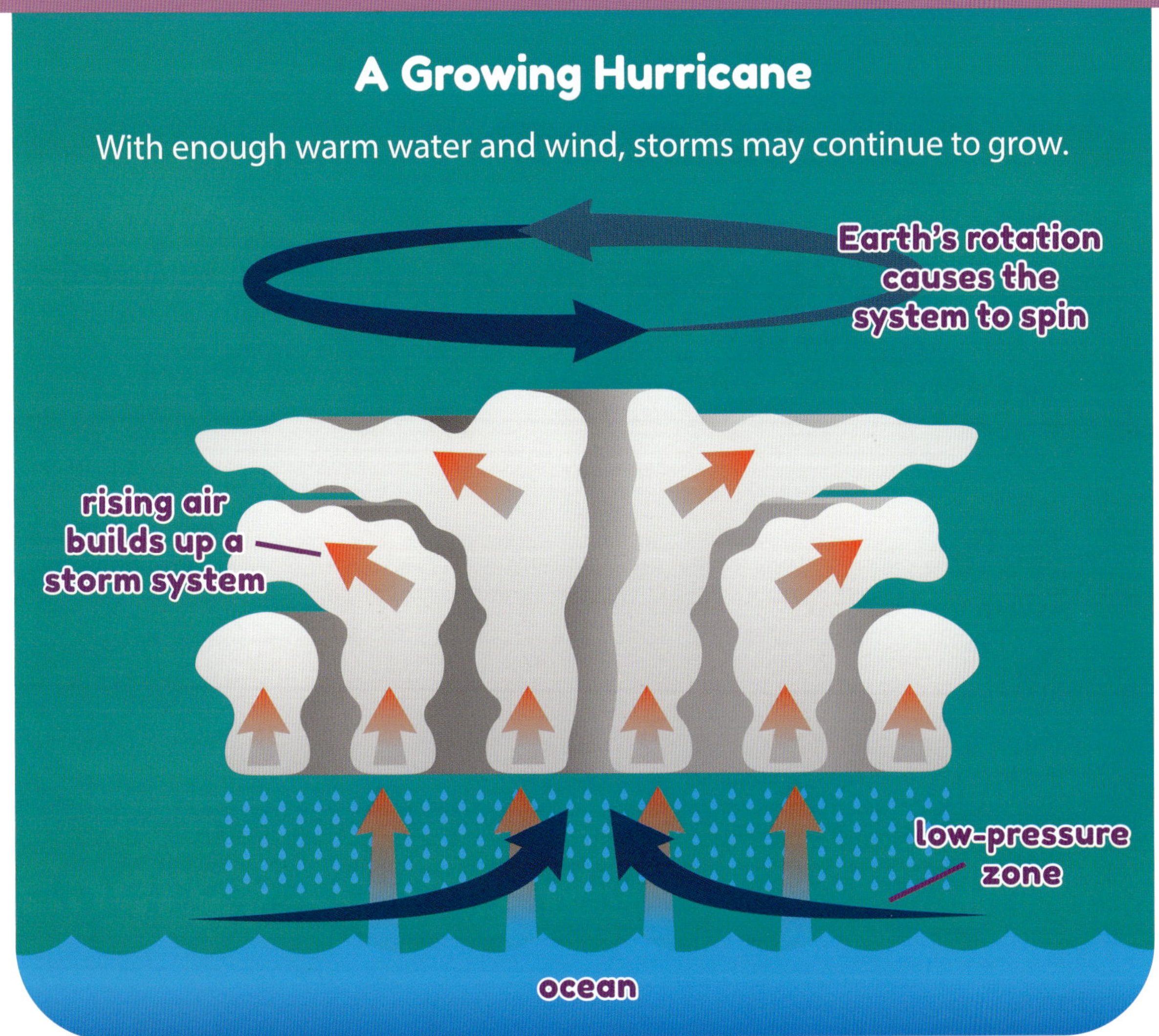

Energy to Grow

Warm ocean water and humid air give a tropical disturbance the energy it needs to grow. The storm's winds move faster. The clouds formed by cooling air grow bigger. They stretch higher.

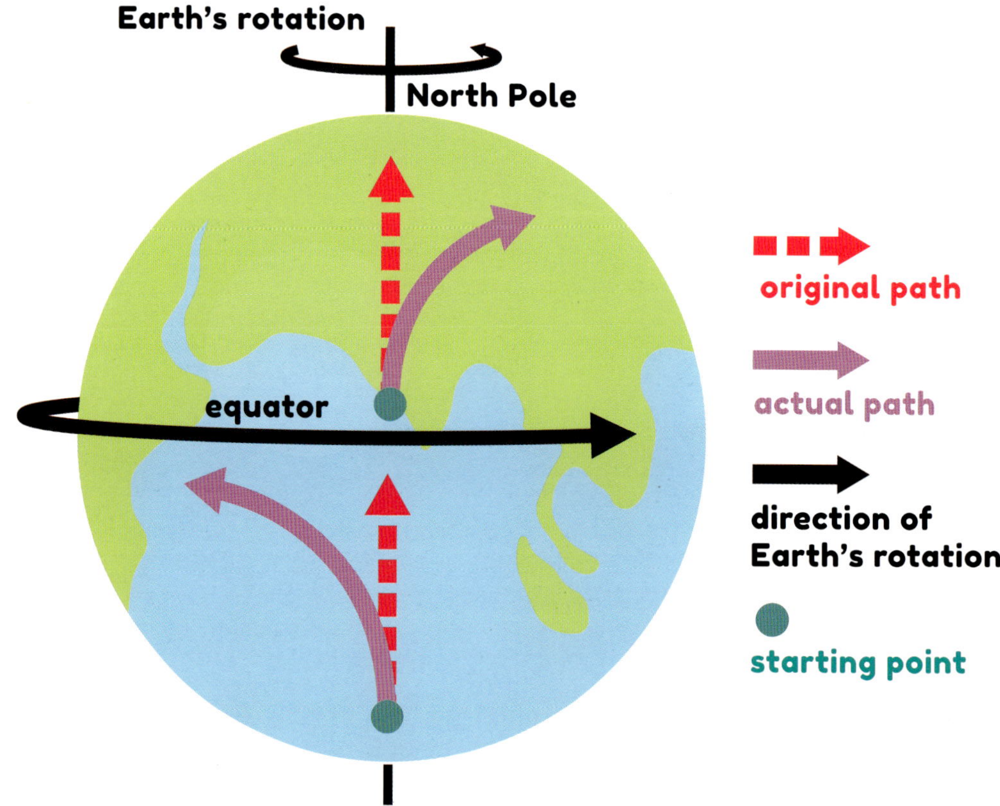

Earth's rotation causes winds to curve in opposite directions north and south of the equator.

The Coriolis Effect

Earth's rotation causes wind to curve. This curving is called the Coriolis effect. North of the equator, air pulled into the storm's center curves to the east. This makes the storm rotate counterclockwise. South of the equator, the air curves to the west. This makes the storm rotate clockwise.

Tropical Depressions

Big storms start experiencing the Coriolis effect. These storms become known as tropical depressions. Tropical depressions have wind speeds of less than 39 miles per hour (63 kmh).

Some tropical depressions are large enough to be seen from space.

Tropical Storms

Some storms get stronger. They become tropical storms. Tropical storms have wind speeds of 39 to 73 miles per hour (63 to 117 kmh).

Ocean storms may grow for hours or days before becoming hurricanes.

The strongest recorded hurricane wind speed was 215 miles per hour (346 kmh).

Hurricanes

Sometimes tropical storms get even stronger. They become hurricanes. Hurricanes have wind speeds of 74 miles per hour (119 kmh) or more. These are the strongest types of tropical storm systems.

The eye is warmer than the rest of the hurricane.

Eye of the Storm

The eye of a hurricane forms when a storm's winds reach about 74 miles per hour (119 kmh). The eye is usually 20 to 40 miles (32 to 64 km) across. The storm's air rotates around the eye. The weather in the eye is normally calm. It has little rain or wind.

Storms of Many Sizes

Hurricanes come in different sizes. A typical hurricane is about 300 miles (480 km) wide. But some hurricanes stretch more than 1,000 miles (1,600 km).

The largest hurricane ever recorded was about 1,380 miles (2,220 km) wide.

Engineer Herbert Saffir and meteorologist Robert Simpson, *pictured*, created the Saffir-Simpson Hurricane Wind Scale in the early 1970s.

Saffir-Simpson Hurricane Wind Scale

Scientists classify hurricanes by wind speeds. They use the Saffir-Simpson Hurricane Wind Scale. It measures a hurricane's strength. It also measures the storm's potential to cause damage.

Rating Hurricanes

The Saffir-Simpson Hurricane Wind Scale rates each hurricane from 1 to 5. Category 1 and 2 hurricanes are the weakest. But they are still very dangerous. They have winds of 74 to 110 miles per hour (119 to 177 kmh). These winds can uproot trees and damage roofs.

Saffir-Simpson Hurricane Wind Scale

The Saffir-Simpson Hurricane Wind Scale helps scientists measure how dangerous a hurricane is.

CATEGORY	WIND SPEED (mph)	WIND SPEED (kmh)
1	74–95	119–153
2	96–110	154–177
3	111–130	178–209
4	131–155	210–249
5	156+	250+

Major Hurricanes

Hurricanes in Categories 3 through 5 are considered major hurricanes. These hurricanes are devastating. They can damage or destroy houses. Hurricanes can cause power outages lasting days or weeks.

In 2023, Hurricane Idalia hit Florida as a Category 3 hurricane.

Hurricane Michael hit Florida in 2018 as a Category 5 hurricane.

The Strongest Hurricanes

The strongest hurricanes are Category 5 storms. These hurricanes have winds of 156 miles per hour (250 kmh) or more. These winds are powerful enough to destroy buildings.

As hurricanes weaken, they become less symmetrical.

The Storm Weakens

Hurricanes weaken as they travel over cooler ocean water. There is less energy and humid air to fuel the storm. Hurricanes also weaken as they move over land. They are no longer getting energy from warm ocean waters. But even a weakened hurricane can cause great damage.

Death of a Hurricane

Hurricanes weaken as they run out of energy. Some break apart. Others are absorbed into other storm systems. Still others turn into different types of storms.

In 2022, Hurricane Ian broke apart after losing its energy over land.

People who want to become weather scientists can earn degrees in meteorology.

Scientists at Work

Scientists can predict when a storm will become a hurricane. They can predict the hurricane's path. They can predict when and where it will hit land. They can also forecast how severe it will be.

Saving Lives

Hurricane predictions save many lives. Scientists can warn anyone who is in the path of a dangerous hurricane. Then those people can flee or prepare for the storm.

The first known person to predict hurricanes was a Cuban priest named Benito Viñes. He began predicting hurricanes in the 1870s.

FUN FACT!

Computers can use artificial intelligence to forecast hurricanes.

Hurricane Season

Hurricane season is the time of year when hurricanes are most likely to form. Hurricane season runs from June 1 to November 30 in the Atlantic and Central Pacific Oceans. Hurricane season is from May 15 through November 30 in the Eastern Pacific Ocean.

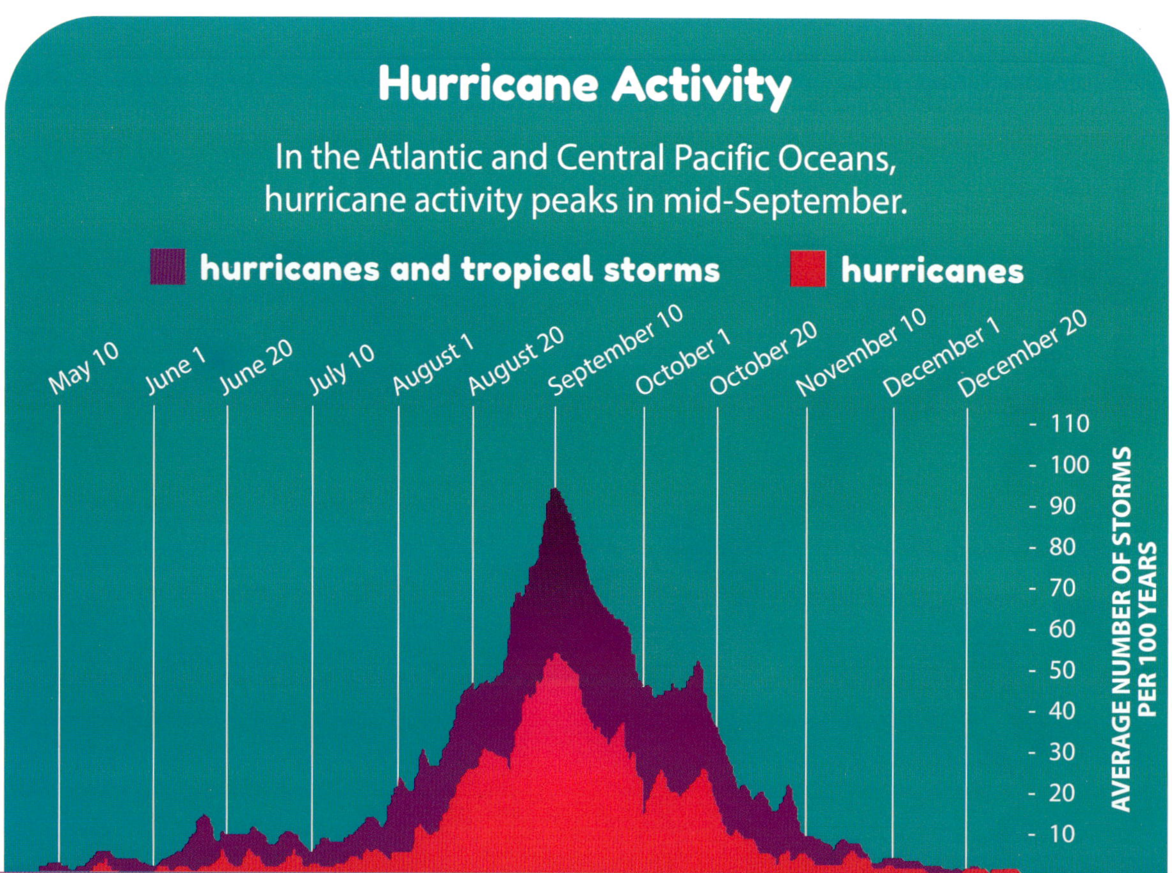

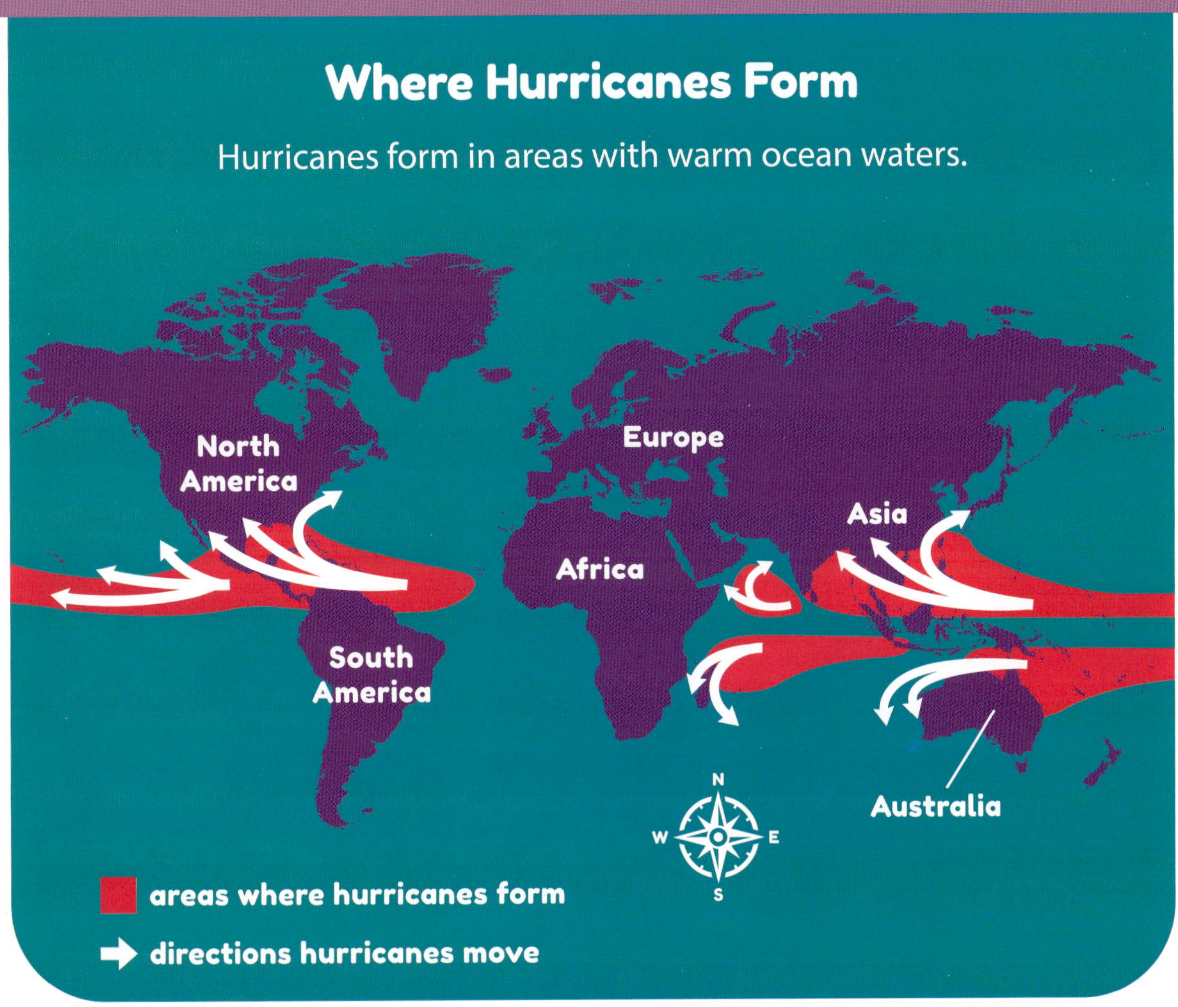

Decades of Data

Scientists base hurricane season estimates on the data they have collected over the last several decades. Looking at the past can help them predict the future. They share this data with the public to keep people informed.

About 45 percent of tropical storms become hurricanes.

Named Storms

Storms are named when they become tropical storms. Named storms have sustained winds of at least 39 miles per hour (63 kmh).

A to Z

The World Meteorological Organization keeps a list of storm names. The list is in alphabetical order. The first named storm of a season receives a name that starts with *A*. The next storm gets a name that starts with *B*. Naming storms helps people keep track of them.

Retired Names

The World Meteorological Organization retires a hurricane name if the storm kills many people or is very destructive. Katrina and Ian are retired names.

The headquarters for the World Meteorological Organization is in Switzerland.

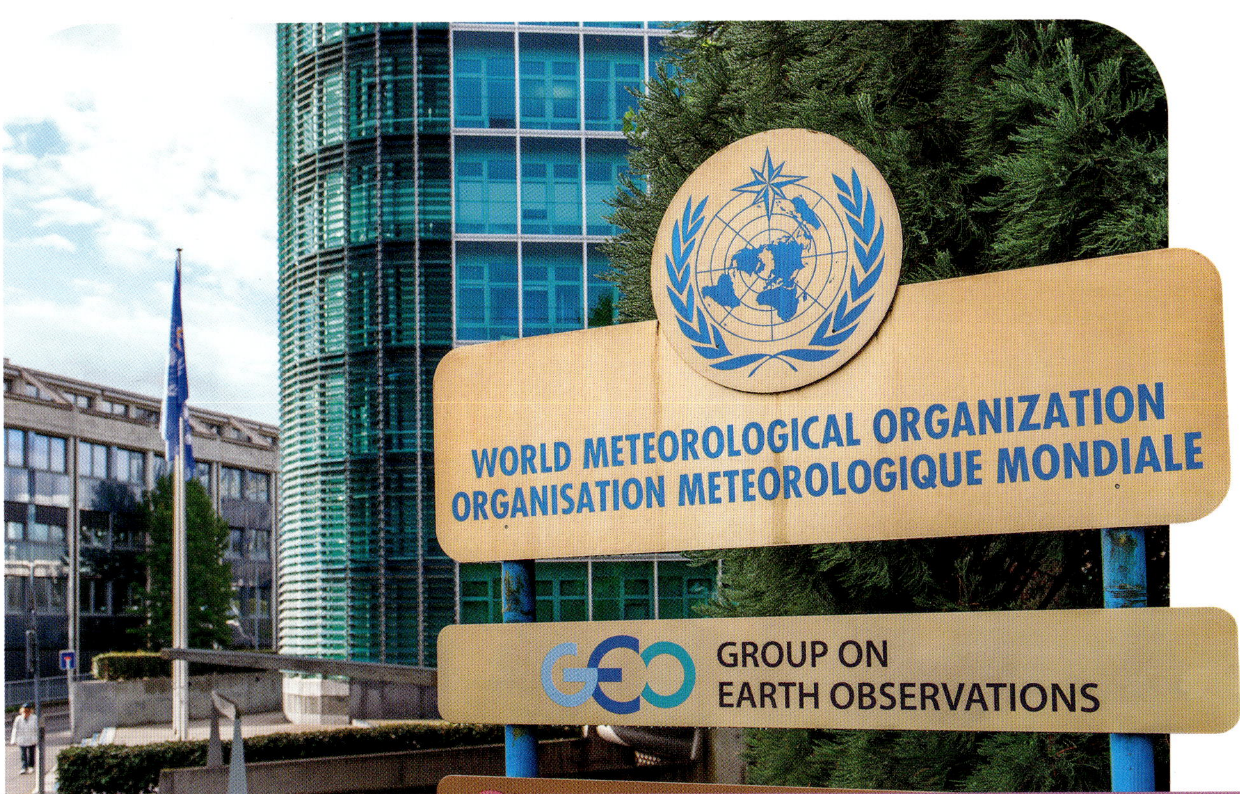

Gathering Data

Scientists use data to predict hurricanes. They gather data about water temperatures. They also gather data about wind speed and direction. Scientists use this information in computer models.

Even with advanced tools and computer models, meteorologists' predictions are not always accurate.

Scientists use many different types of computer models to predict and track hurricanes.

Computer Models

Computer models predict when and where hurricanes are most likely to form. They predict how intense hurricanes will be. This helps scientists forecast and track hurricanes.

Weather satellites can take pictures of storms every 30 seconds.

Satellites

Scientists use several tools to gather data. These tools include satellites. Thousands of satellites circle Earth in space. Some satellites track the weather. They spot storms forming far out at sea. Satellites watch the storms grow and strengthen. They take images of the storms.

Monitoring Storms

Scientists look at pictures from satellites. They use the images to predict how big the storm will get. The images also help scientists forecast where the storm will move.

An Average Season

The Atlantic Ocean averages about ten tropical storms each hurricane season. Six of those storms usually become hurricanes. On average, two will grow to Category 3 or greater.

The National Oceanic and Atmospheric Administration (NOAA) puts some of its satellite images online to let people around the world track storms.

Infrared Sensors

Many satellites have infrared sensors. These sensors use light and heat to detect objects. They produce images that show objects in bright colors. This can help scientists track hurricanes at night.

The colors in infrared images represent the temperature of different parts of an object. Warmer areas are orange and red. Cooler areas are blue and green.

The shape of a hurricane's eye changes as a storm strengthens and weakens.

Tracking a Hurricane's Eye

Infrared sensors can show the size and shape of a hurricane's eye. This can help scientists determine how strong a storm is. Hurricanes with clearly defined eyes are strong. Hurricanes with undefined eyes are weaker.

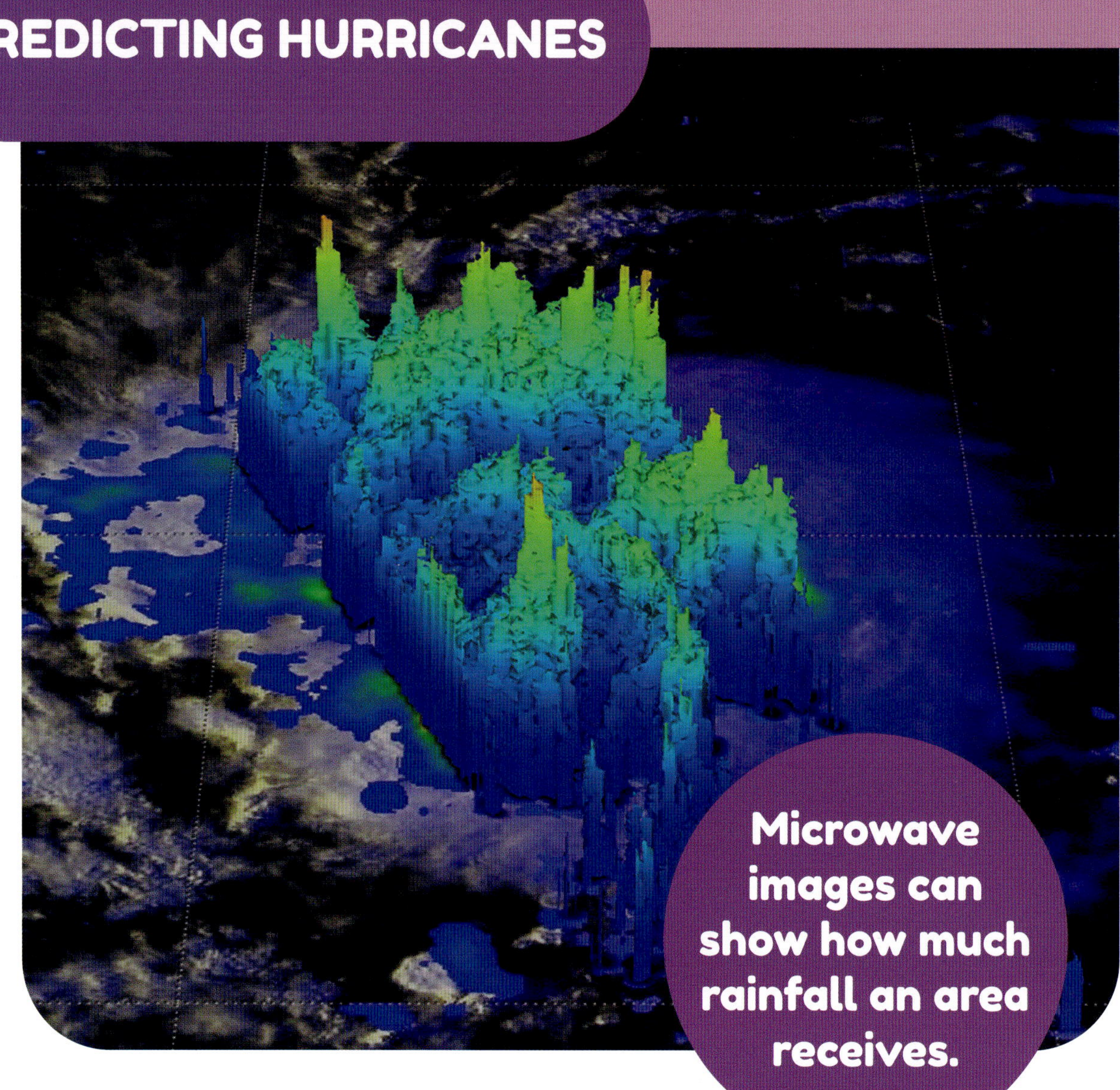

Microwave images can show how much rainfall an area receives.

Microwave Images

Some satellites use microwave technology to look inside storm clouds. This technology is like an X-ray. Scientists use it to look inside a hurricane. This lets them see the parts of the storm.

Showing a Storm's Intensity

Microwave images help scientists determine how intense a hurricane is. These tools help scientists determine if a storm is getting stronger or weaker. Microwave images can also help scientists see hurricanes through clouds.

The Global Precipitation Measurement (GPM) satellite has microwave sensors.

Aircraft

Aircraft collect hurricane data. Trained pilots fly planes into storms to get information. Sometimes they fly planes into a hurricane's eye.

FUN FACT!

In 1943, a pilot flew an airplane into a hurricane for the first time. The scientist onboard collected data from inside the hurricane.

The United States has a team of pilots trained to fly planes into hurricanes. These pilots call themselves the Hurricane Hunters.

Aircraft traveling into hurricanes have crews of five people. Each crew has a pilot, a copilot, a navigator, a meteorologist, and a person in charge of collecting weather data.

Aircraft Instruments

Aircraft carry weather instruments such as radar and sensors. These tools measure wind speeds and air pressure. They can also measure air temperature and humidity. The plane's crew collects data about the storm throughout the flight.

Dropsondes are loaded into launcher tubes in aircraft and shot into storms.

Dropsondes

Aircraft release dropsondes near or in hurricanes. A dropsonde is a tube that carries scientific instruments. It floats down with a parachute.

Collecting Data

Dropsondes collect data about a storm as they fall. They record the storm's temperature and humidity. They also record wind speed and direction. They record the surrounding pressure. Some dropsondes collect data from the ocean. The dropsondes send their data to the aircraft.

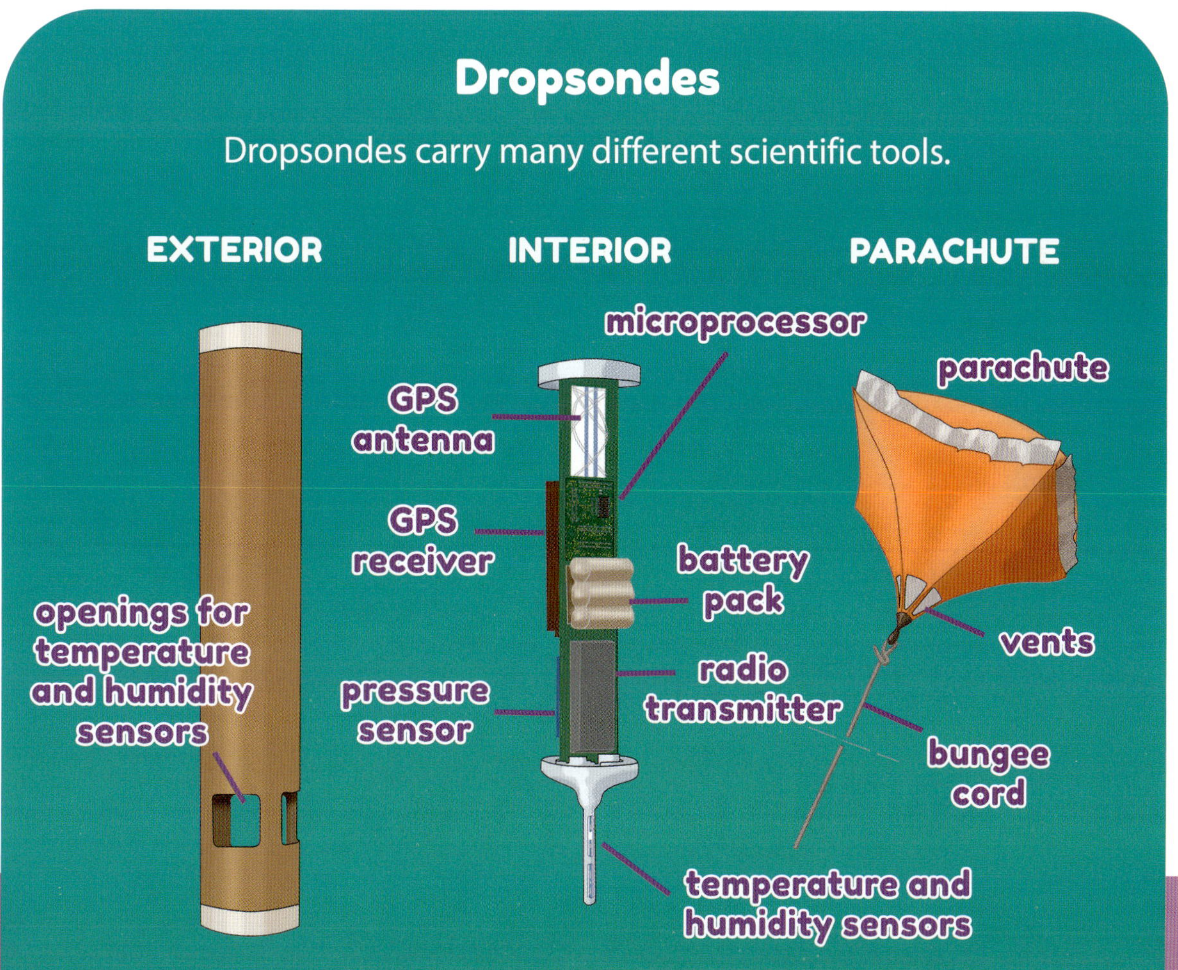

The Altius-600 is a drone that scientists can control remotely.

Aerial Drones

Pilotless aerial drones also fly into hurricanes. This allows scientists to collect data without risking human lives. Some drones can fly for 18 hours or longer.

Probes

Crew members drop probes from ships. The probes measure water temperature as they sink below the surface. They send measurements back to ships using small wires.

NOAA is part of an international group that gathers weather data to help scientists predict hurricanes.

Ships

Ships carry sensors. The sensors take air pressure and temperature measurements. They record the sea surface temperature. They track wind speed and direction. Ships' crews send the data to weather stations on land.

Weather Buoys

Weather buoys are big platforms with sensors. They float on the ocean surface. Scientists use weather buoys to monitor areas where hurricanes form.

The United States manages more than 1,300 weather buoys.

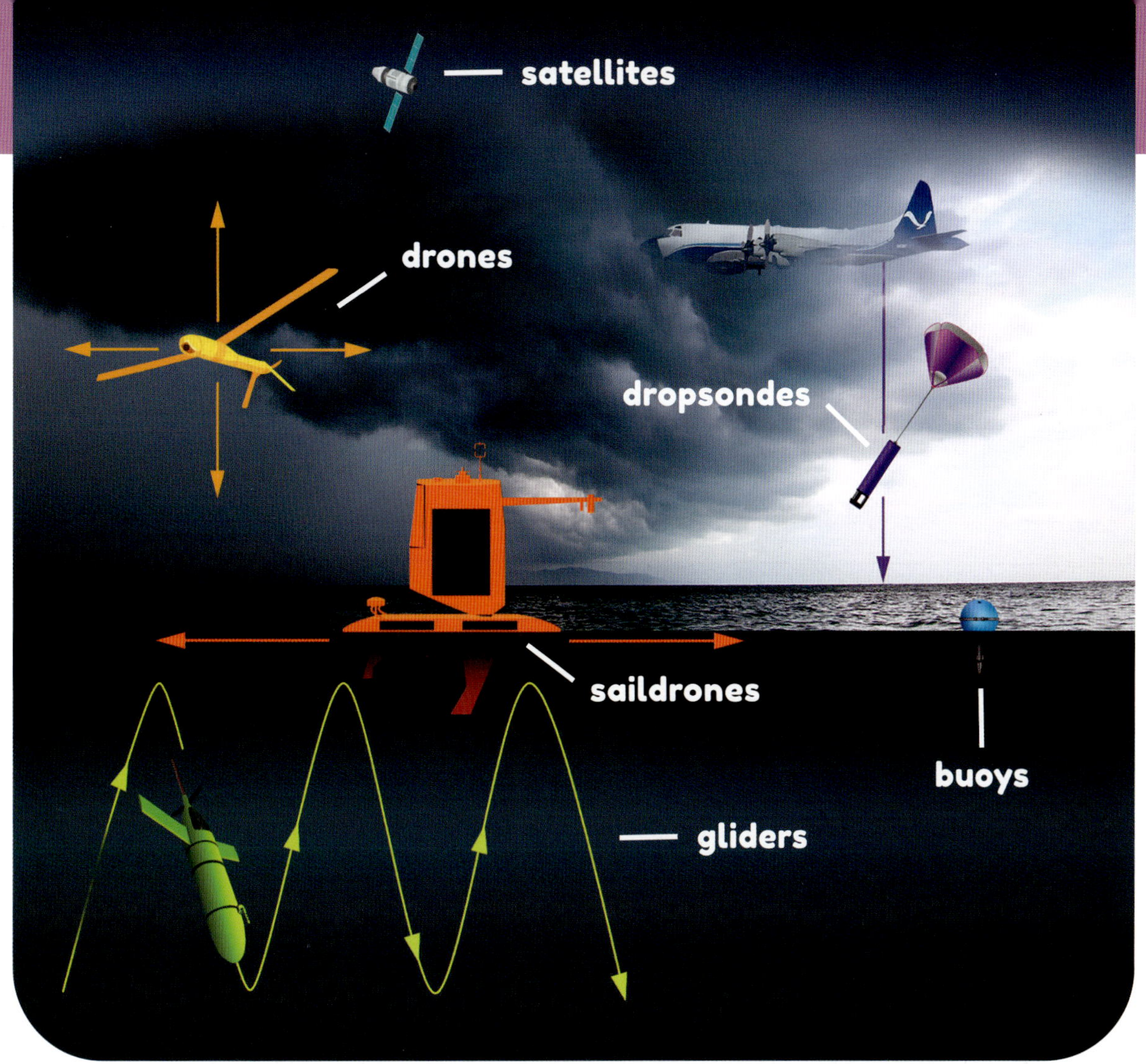

Scientists use many tools to gather data about hurricanes.

Data from the Ocean

Buoys measure wave height and speed. They measure air and ocean temperatures. They also measure wind speed and direction. This information can help predict hurricanes.

About 40 percent of the hurricanes that reach the United States hit Florida. Hurricane Irma hit Florida in 2017.

Wind and Water

Hurricanes are powerful storms. They bring strong winds and heavy rains. They also cause storm surge. They do major damage to coasts and cities. Hurricanes put many lives at risk.

Dangerous Storms

Hurricanes can hurt people and damage property. Debris blown by the wind can hurt people. Flooding injures people too.

FUN FACT!

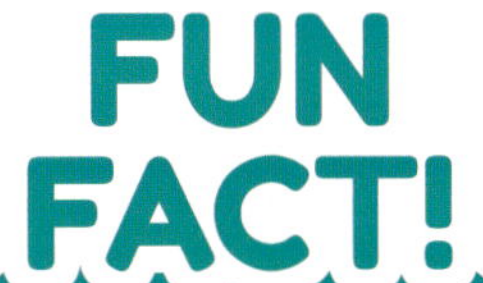

A major hurricane releases the energy of ten atomic bombs every second.

Strong winds can fling branches, furniture, and even cars.

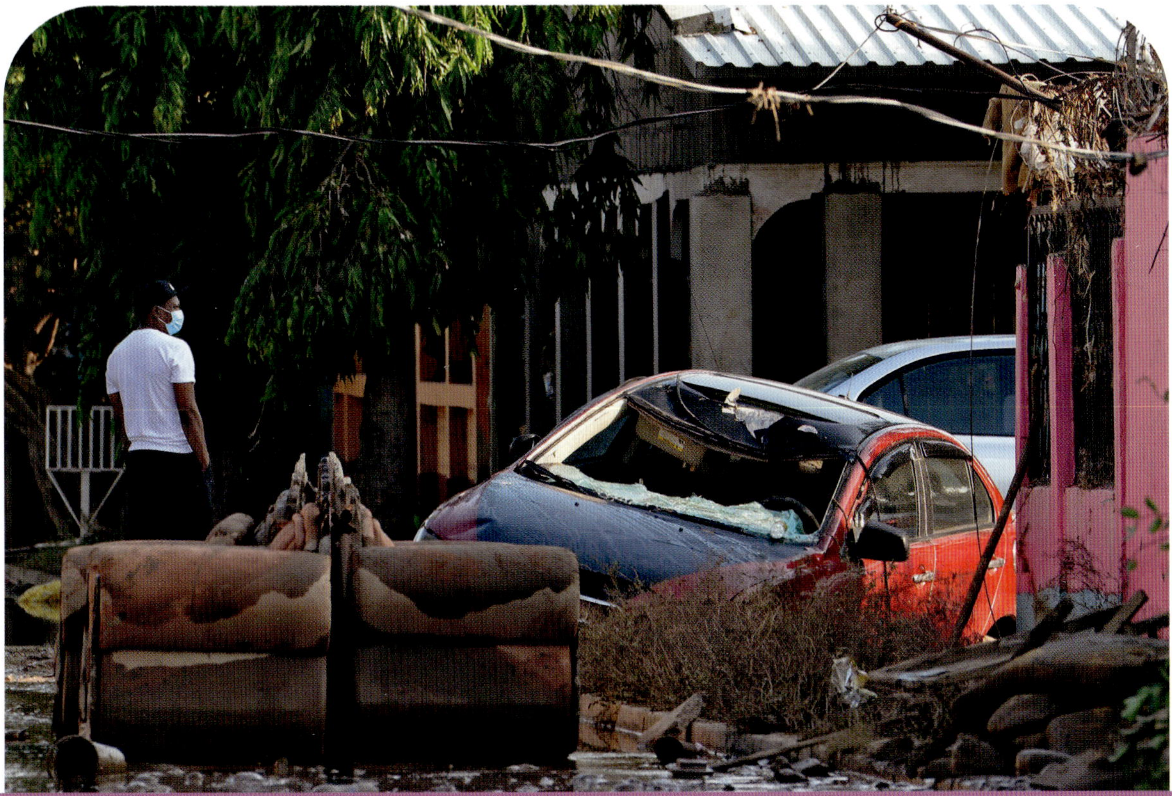

High Winds

Hurricane winds are mighty. They can knock down power lines. They can rip off roofs. They overturn cars. They can even destroy buildings.

Hurricane Ian damaged more than 100,000 homes.

Only four Category 5 hurricanes made landfall in the United States between 1935 and 2023.

Landfall

Hurricanes are especially dangerous when they make landfall. Landfall is when the eye of a hurricane hits land. This is dangerous because hurricane winds are strongest near the storm's eyewall. But the winds can still do damage many miles out. The stronger and faster a hurricane is, the farther its powerful winds will reach.

A hurricane can release more than 2.4 trillion gallons (9 trillion L) of rain every day.

Inland Rain

Hurricanes gather a lot of moisture over the ocean. They drop enormous amounts of rain as they move over land. The rain can fall many miles inland.

Inland Flooding

Flooding can happen when a lot of rain falls over a short time. The rain causes lakes and rivers to rise. These bodies of water can overflow if their water levels get too high.

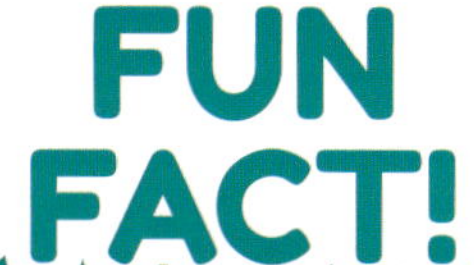

The average hurricane covers land in six to 12 inches (15 to 30 cm) of rain.

Flooding can become so bad after a hurricane that people may travel in boats or kayaks.

Dangerous and Deadly

Flooding damages property. It also can be deadly. From 2013 to 2022, freshwater flooding caused 57 percent of hurricane deaths. Flooding can last for weeks after a hurricane.

Storm Surge

A hurricane's powerful winds push large amounts of water toward shore. This causes the sea level

Even weak hurricanes can cause disastrous flooding.

to rise along the coast. This rise is called storm surge. The amount of storm surge depends on a hurricane's strength, size, and speed. Storm surge can cause dangerous flooding. It can raise the sea level by more than 30 feet (9 m). From 2013 to 2022, 11 percent of hurricane deaths were caused by storm surge.

Category 5 hurricanes can cause a storm surge of more than 19 feet (5.8 m).

Flooding the Coast

Storm surge can flood coastal lands. This flooding can damage coastal buildings. It can wash away roads. Floodwaters can also damage bridges. A strong storm surge can leave coastal areas uninhabitable for a long time.

FUN FACT!

The largest storm surge on record occurred in Australia in 1899. Sea levels rose by 43 feet (13 m).

Tornadoes

Storm clouds can form tornadoes when hurricanes hit land. A tornado is a rotating column of air. It stretches from the bottom of a cloud to the ground. A tornado's strong rotating winds pick up debris. The debris caught in the wind makes the tornado look like a spinning funnel cloud.

It is difficult for meteorologists to predict if hurricanes will form tornadoes.

Tornadoes Add Damage

Tornadoes usually form in a hurricane's rainbands. They occasionally form in the eyewall. These tornadoes are usually weak compared with tornadoes that do not form in hurricanes. But they can still be destructive.

Hurricane Ivan hit the United States in 2004. It created 120 tornadoes.

Florida's mangrove trees were destroyed by Hurricane Andrew's strong winds. The mangroves still have not fully recovered.

Damage from Hurricane Andrew

Hurricanes have a significant effect on the environment. Hurricane Andrew struck Florida in 1992. The storm damaged Florida's coastal forests. Almost 100 percent of the area's mangrove trees were damaged. These forests are important for preventing erosion. Their destruction puts the area in more danger from future hurricanes.

Dunes are hills made of sand. They can block waves and storm surge from reaching homes.

Coastal Erosion

Storm surge can move coastal sand. This reshapes the shoreline. Hurricane winds and storm surge erode beaches. Dunes are destroyed. Marshes and cliffs erode. These changes get rid of natural hurricane barriers. This can leave coastal areas less protected from future storms.

Beach Erosion in Florida

Hurricane Idalia struck Florida in 2023. The hurricane caused erosion along every beach in Pinellas County. The storm surge pulled sand away from the beaches. Many places had little beach left.

Salt Marshes

Salt marshes are wetlands near the coast that absorb water. They act as a natural buffer to protect the coast from storm surge.

Some coastal areas are given extra sand to protect beaches from hurricanes.

Habitat Damage

Hurricanes damage the places where animals and plants live. Strong winds and flooding can kill animals. The wind and water can also make food hard to find.

Harming Wildlife

Hurricane Hugo struck Puerto Rico in 1989. It damaged

Even fish can be killed by hurricanes.

Millions of Puerto Rican parrots used to live in the wild. In 2024, fewer than 50 were left.

many forests. Birds and bats that ate fruits from the forests could not find food. Many died. One bird species affected was the Puerto Rican parrot. The parrot lost half its population after Hurricane Hugo.

FUN FACT!

Some scientists believe that birds can tell when hurricanes are coming.

Millions of farm animals are killed in hurricanes.

Threat to Farms

Hurricanes can also hurt farms. They damage crops and kill livestock. They can make soil less suitable for growing. Hurricane Mitch hit Honduras in 1998. The hurricane destroyed 70 percent of the country's crops.

Fruit Trees at Risk

Tropical fruit trees are vulnerable to hurricanes. Powerful winds rip leaves off fruit trees. They knock fruit to the ground. This makes the fruit rot. Hurricane winds can also snap fruit tree branches. They may also rip entire trees from the ground.

Hurricane Ian caused $675 million of damage to Florida's citrus trees.

Steps to Safety

Hurricanes can be deadly. They can affect those on the coast and farther inland. When a hurricane approaches, the people in its path can take several steps to stay safe.

Hurricanes are the most devastating natural disasters in the United States.

People can follow simple steps to prepare for hurricanes.

Understand the Risks

Everyone should understand the risks of a hurricane. People should know if they live in a place where storm surge will be a problem. They should know if they have a lot of windows that high winds could damage. They should be aware of nearby rivers that could flood. Understanding these risks is essential to preparing for a hurricane.

Make a Plan

Everyone should have an emergency plan. They should keep emergency numbers in a safe place. They should decide on a place to seek shelter. People should know the evacuation route if officials order everyone to leave.

Some cities have signs to mark evacuation routes.

Stores may run out of food and water before storms, so people should stock up on supplies ahead of time.

Gather Food and Water

Roads may be blocked or flooded during hurricanes. Power and water may not be available. People should prepare by gathering necessary supplies before a storm. They should make sure they have enough food and water.

Emergency Supply Kit

Everyone should have an emergency supply kit ready, even if they do not live in an area that gets hurricanes.

Other Supplies

Other helpful emergency supplies include a battery-powered radio and a flashlight. Kits should also have extra batteries. A first aid kit can be helpful if anyone is injured in a hurricane. Some people pack a whistle to signal for help.

Prepare the Home

Many people prepare their homes before a storm to reduce property damage. They move outdoor furniture and other objects inside. This prevents items from blowing away.

Objects left outside could break or get caught in the wind and hurt people.

Cover Windows and Turn Off Power

Many people cover windows and doors with storm shutters or wood. Covering windows and doors can protect them from damage. Some people also turn off power in case of flooding. This can prevent water from carrying electric currents that may cause electric shocks.

Windows should be boarded up at least 48 hours before a hurricane is expected to arrive.

NOAA and the Federal Emergency Management Agency work together to release weather emergency alerts.

Emergency Alerts

During a hurricane, emergency officials release information about the hurricane's path and strength. Emergency alerts warn about flooding and road closures. They may also give information about power outages. These alerts help communities stay safe.

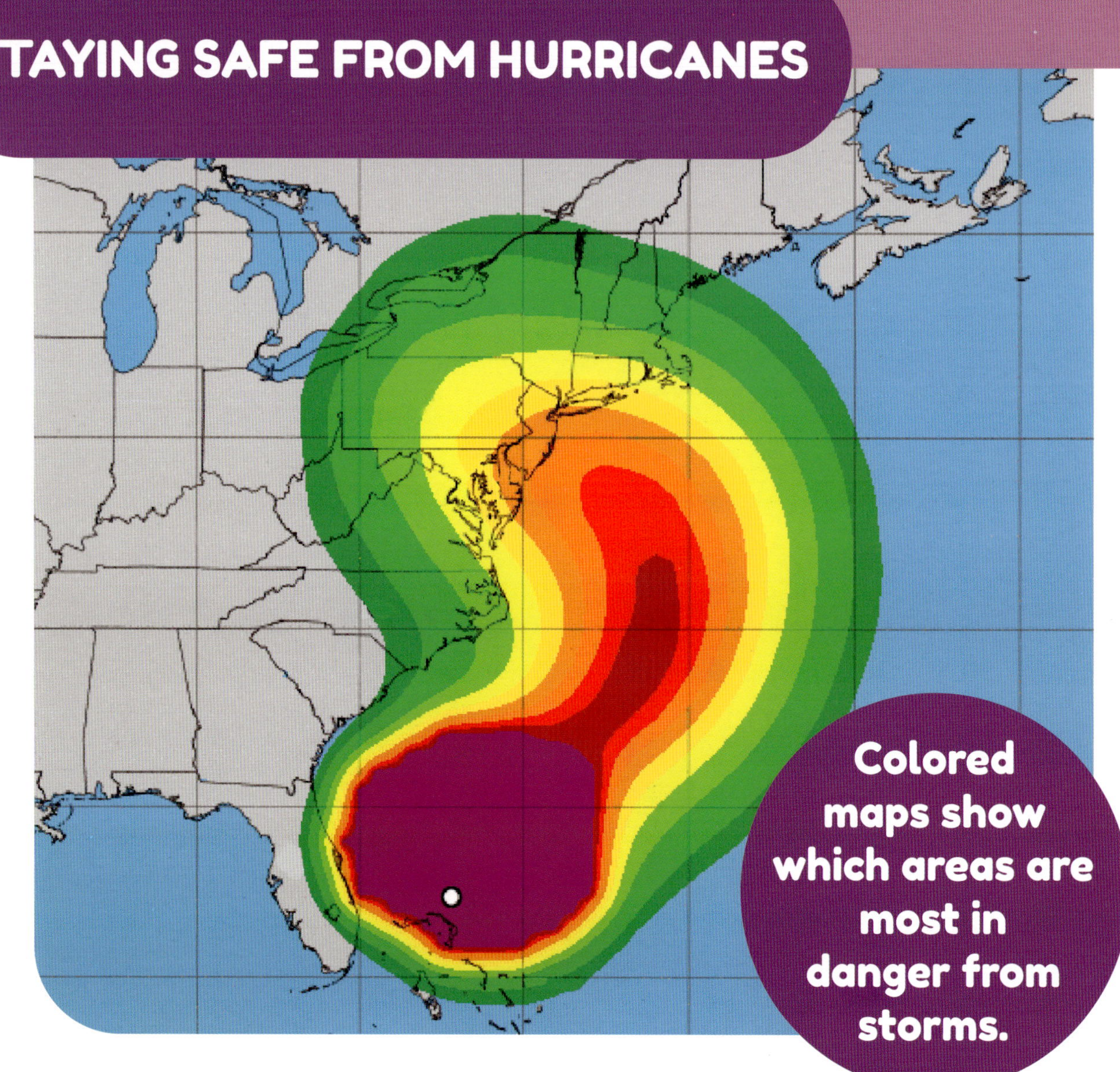

Colored maps show which areas are most in danger from storms.

Hurricane Watch

One alert that the National Weather Service issues is a hurricane watch. This means that a hurricane has the potential to form within 48 hours. It alerts people at risk.

Hurricane Warning

Sometimes the National Weather Service issues a hurricane warning. This warning tells people a hurricane is expected to hit within 24 hours. This gives people time to prepare or evacuate.

The National Weather Service sends out emergency alerts to phones and smart devices.

Ready to Go

People should fill their vehicles' fuel tanks before a hurricane. They should pack emergency bags. They must prepare to leave at a moment's notice.

Lines for gas are often very long before and after hurricanes.

It is illegal to stay home when the government orders a mandatory evacuation.

Mandatory Evacuation

Sometimes a hurricane becomes so dangerous that it is not safe to stay in the storm's path. Government officials may order a mandatory evacuation. This means that everyone in an area has to leave. Mandatory evacuations save lives.

The lowest level of a home is safest from high winds.

Hurricane at Home

Sometimes the government tells people to stay at home during a hurricane. A storm shelter or interior room is the safest place to shelter from a hurricane. These areas should be away from windows or glass doors that may shatter.

Safe Shelters

People who live in temporary buildings or mobile homes should find a sturdy building to shelter in. Some coastal areas have designated hurricane shelters. These may include schools or convention centers.

Emergency evacuation shelters may have to house thousands of people.

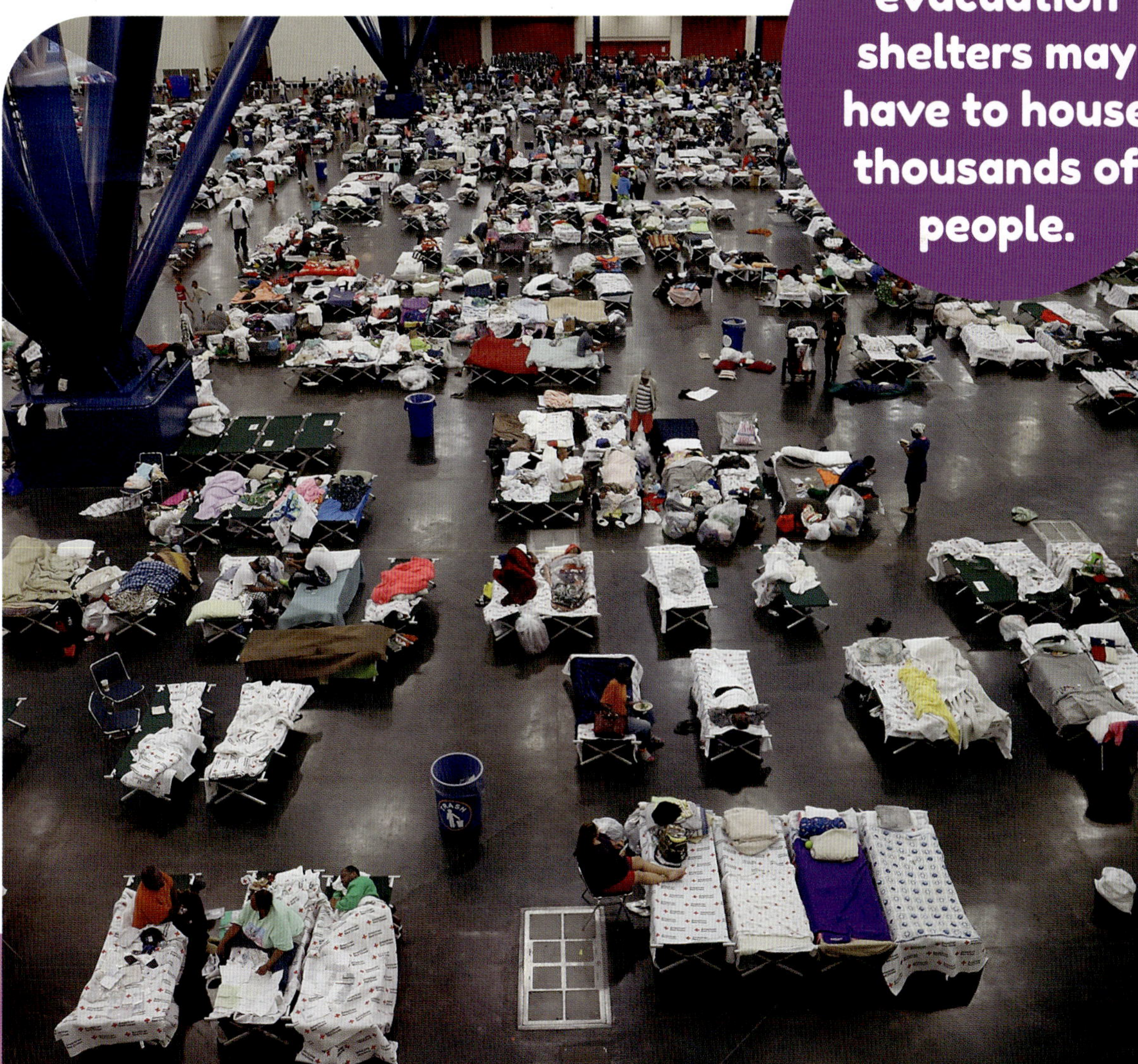

Find High Ground

People should find high ground if they see flooding. Some people climb to the highest part of their house. But it is important to have a way to exit the building if the water keeps rising.

Flood Maps

The Federal Emergency Management Agency (FEMA) makes flood maps. These maps show which areas have a higher risk of flooding during a storm.

People can get stranded on their roofs during hurricanes.

Cars do not protect people from floodwaters.

Rising Floodwaters

People should stay away from flooding. Trying to drive or walk through floodwaters is very dangerous. Many drown attempting this.

During Hurricane Harvey, one-third of the city of Houston was underwater.

Fast-Moving Water

Floodwaters can move quickly. As little as six inches (15 cm) of floodwater can knock people off their feet. Twelve inches (30 cm) of moving floodwater can carry a car away.

Inside the Eye

There may be a short period of calm during a hurricane. This happens when a hurricane's eye passes over an area. People should not go outside during this time. They should stay in their shelters.

It can take hours for a hurricane's eye to pass over an area.

After the Storm

Those who evacuate should not immediately return after a hurricane. Roads may be blocked. Bridges may be washed out. Some areas may be flooded. Local officials will announce when it is safe to return.

Power might be out for months after a hurricane.

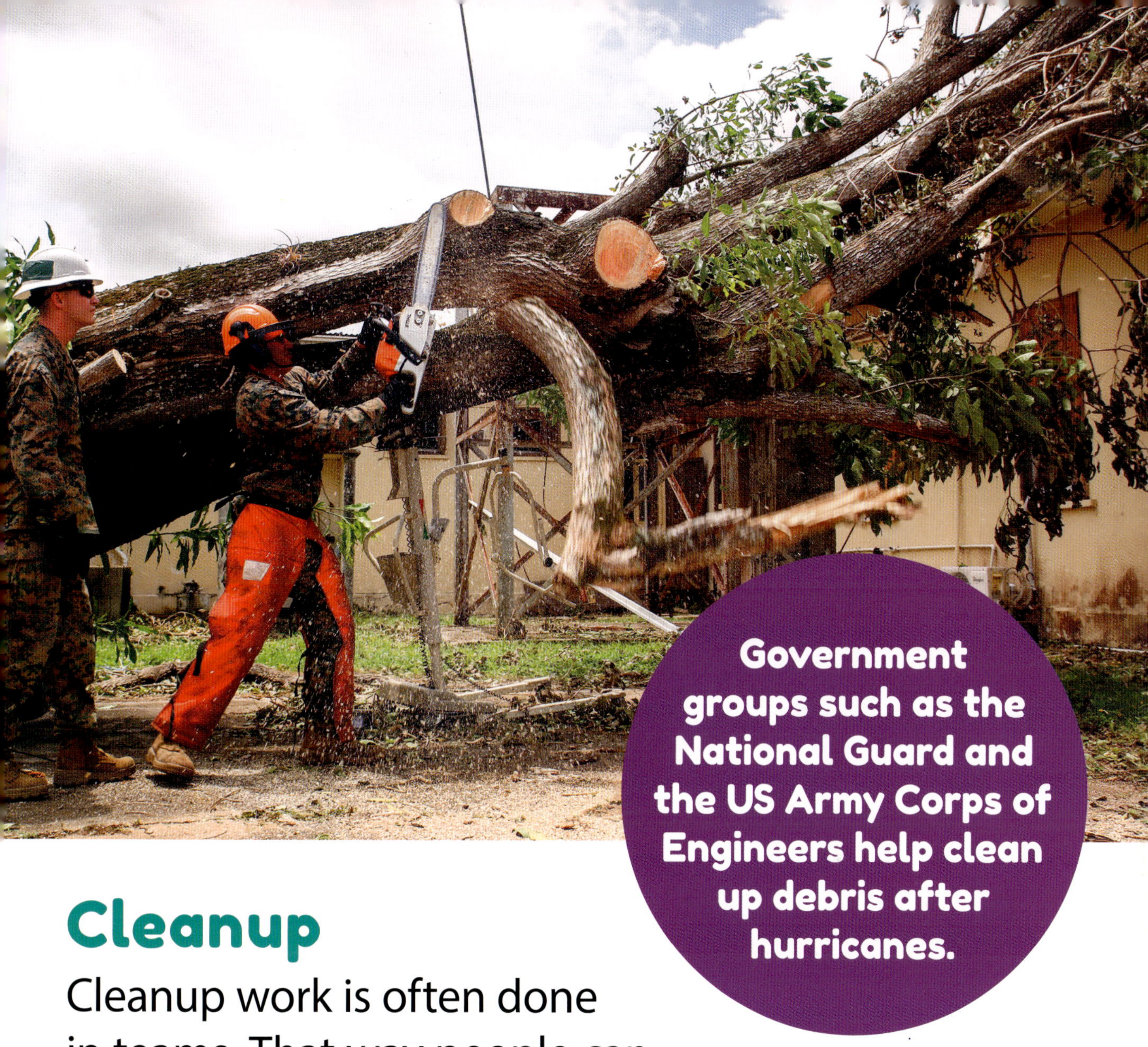

Government groups such as the National Guard and the US Army Corps of Engineers help clean up debris after hurricanes.

Cleanup

Cleanup work is often done in teams. That way people can help if someone gets hurt. Everyone on a team should wear protective clothing when cleaning up damage. Protective clothing includes gloves and boots. Team members should avoid all wet electrical equipment to prevent shocks.

If people have to walk in hurricane floodwater, they should wear waterproof boots.

Avoid the Water

People should avoid walking in floodwater after a hurricane. The water is often full of dangerous debris. It may also contain chemicals and waste. The water might carry germs that can make a person sick. Fallen electric power lines can also make floodwaters unsafe.

Animals after a Hurricane

A hurricane may carry wild animals to places they are not normally found. Animals such as rats and snakes may be trapped. Moving carefully during cleanup helps people avoid being bitten.

In 2023, Hurricane Idalia struck Florida and carried flamingos as far away as Wisconsin.

A Changing Planet

Earth's climate is changing. Climate change is a change in an area's usual weather patterns. Some climate change is normal. Earth's climate has changed over its history. These changes have happened naturally in the past.

Climate change is affecting animals around the world.

From 1880 to 2020, Earth's average temperature increased by 2 degrees Fahrenheit (1°C).

Rising Temperatures

Earth is getting warmer much faster than it did in the past. This increase in temperature is called global warming. Gases from burning oil and coal trap heat close to Earth. This makes the world hotter.

Climate change makes severe hurricanes, droughts, and even volcanic eruptions occur more frequently.

Extreme Weather

Earth's warming climate is making extreme weather more common. Some extreme weather is normal. Hurricanes are an example of extreme weather. So are tornadoes.

Stronger Storms

The warming climate is making extreme weather more intense. Hurricanes are becoming more powerful. Hurricane winds are stronger. Storm surge is higher. More rain is falling. This is causing hurricanes to become more destructive.

Computer models help scientists predict how global warming will change hurricanes in the future.

Warmer Oceans

Climate change is causing Earth's oceans to warm. Warmer oceans release more water vapor and heat into the air. This provides more fuel for tropical storms. It makes winds stronger. More tropical storms become hurricanes.

More Major Storms

Warmer oceans will fuel more major hurricanes. The number of major hurricanes has increased

More than 90 percent of the extra heat caused by global warming has been absorbed by the oceans.

Climate change is making hurricane winds stronger.

since 1979. At the same time, the number of smaller hurricanes has dropped. In time, more hurricanes will become Category 4 and 5 storms.

A Longer Hurricane Season

Warmer temperatures may cause a longer hurricane season. Warmer oceans allow hurricanes to form earlier in the year. In the Atlantic Ocean, hurricanes are making landfall more than three weeks earlier than they did in 1900.

Hotter temperatures lead to more evaporation. This leads to more water vapor.

More Rain

A warmer atmosphere can hold more water vapor. More water vapor leads to more rain. Future hurricanes are predicted to drop 10 to 15 percent more rain than past storms.

Record Rains

Global warming makes hurricanes release more rain. In 2017, Hurricane Harvey dropped more than 60 inches (152 cm) of rain in some areas. More rain increases the risk of severe flooding.

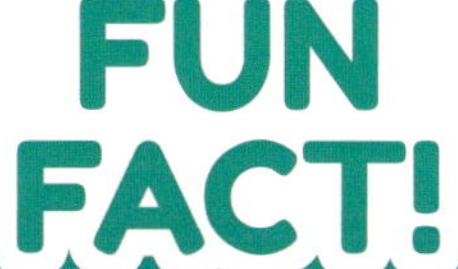

Tropical Storm Imelda brought 44 inches (112 cm) of rain to Texas in 2019.

Scientists estimate that Hurricane Harvey dropped 18 percent more rain on Texas than it would have without climate change.

Rising Sea Levels

Warming temperatures are causing polar ice and glaciers to melt. Melting ice adds more water to Earth's oceans. Warmer ocean temperatures cause seawater to expand. These factors cause sea levels to rise.

FUN FACT!

About 150 billion tons (136 billion metric tons) of Antarctica's ice melts each year.

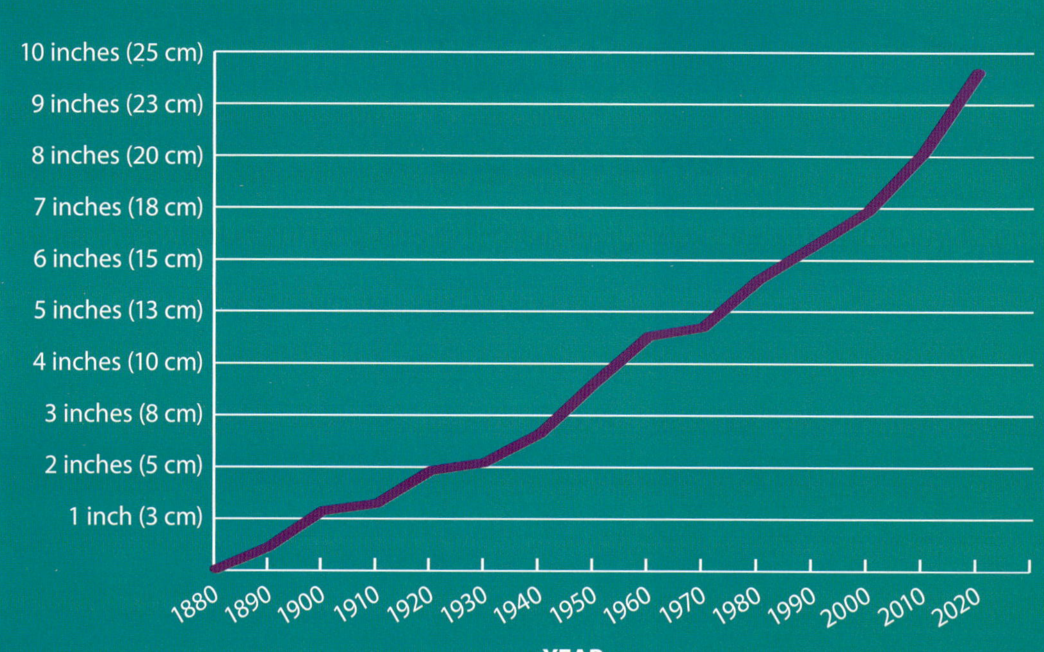

Storm surge will grow more deadly as sea levels rise.

Storm Surge Worsens

Rising sea levels make a hurricane's storm surge worse. They allow storm surge to cause more flooding. This increases property damage. It puts more lives at risk.

When hurricanes grow quickly, people do not have time to evacuate.

Growing Faster

Warming oceans allow hurricanes to grow faster than in the past. In 2023, Hurricane Lee grew from a Category 1 to a Category 5 hurricane in 24 hours. Fast-growing hurricanes are hard to predict. More communities can be caught unprepared for the storm.

Moving More Slowly

Many hurricanes are moving more slowly than in the past. This may be because the climate is becoming warmer. Winds are slower over warmer areas. Slow-moving hurricanes are very dangerous. The longer a storm stays over an area, the more damage it can do.

Slow-moving storms release more rain than fast-moving storms.

New Areas at Risk

Hurricanes cannot form without ocean temperatures of at least 80 degrees Fahrenheit (27°C). Warmer oceans mean there are more places where hurricanes can form. This puts more humans at risk.

Scientists predict climate change will cause more hurricanes to hit the northeastern United States.

Shifting Storms

Tropical storms in the Pacific Ocean have already begun to move north. Similar changes could happen in the Atlantic Ocean in the future. More hurricanes could make landfall in northern regions.

In 2023, climate and weather disasters cost the United States $93 billion.

Increasing Costs

Strong hurricanes do more damage than weaker hurricanes. Cleaning up after them is more expensive. In the United States, four of the most expensive hurricanes in history struck in 2017 and 2018.

Human Health Suffers

Stronger hurricanes put more lives at risk. Illness and disease can quickly spread after a hurricane. But hospitals may be damaged or without power. The sick and injured may not receive the care they need.

Hurricanes kill about 10,000 people every year.

A Growing Storm

Hurricane Katrina began as a tropical depression. The storm strengthened into a hurricane as it moved through the Atlantic Ocean. On August 25, 2005, Katrina made landfall in Florida as a Category 1 hurricane.

Hurricane Katrina was one of the most devastating natural disasters in US history.

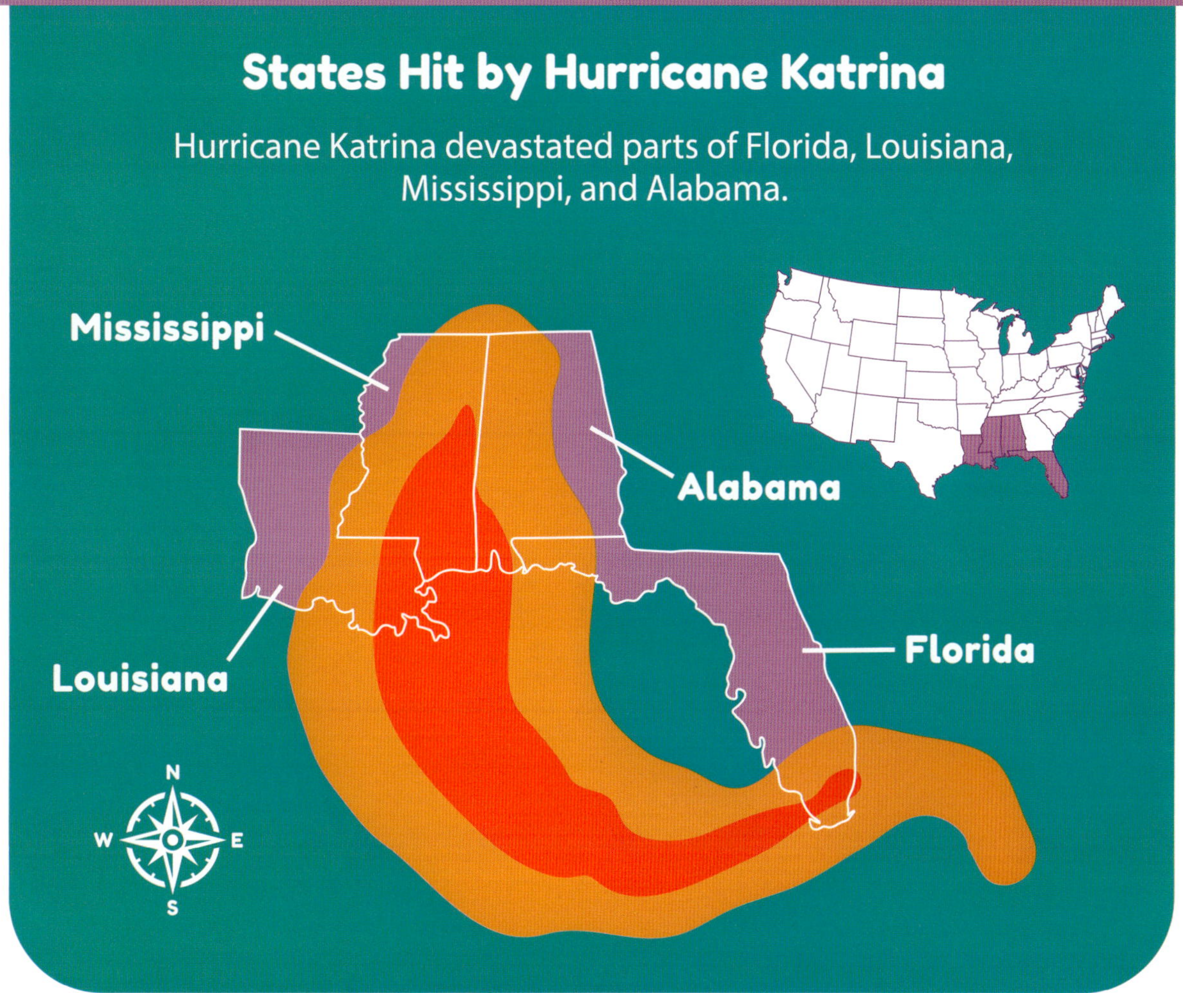

Category 5

Hurricane Katrina moved from Florida to the Gulf of Mexico. The storm grew more powerful over the gulf's warm waters. It reached Category 5 status. The storm's winds were faster than 170 miles per hour (270 kmh).

More than 30,000 people sheltered in the Superdome.

New Orleans Underwater

The mayor of New Orleans issued a mandatory evacuation order. But tens of thousands of people stayed. Some decided to wait out the storm in their homes. Others sheltered in the city's convention center. Many waited in the Superdome football stadium.

Landfall in Louisiana

Hurricane Katrina made landfall in Louisiana on August 29. It had weakened slightly to a Category 4 storm. It hit about 45 miles (72 km) from New Orleans. The hurricane moved along the Mississippi coast.

Rain, wind, and tornadoes battered the states in Katrina's path.

FUN FACT!

When heavy flooding occurs, water can weigh down land and cause an area to sink.

Devastating Storm Surge

Hurricane Katrina's storm surge rose more than 26 feet (8 m). The water washed away buildings. Storm surge eroded beaches. It even swept away an oil rig.

Hurricane Katrina's storm surge washed away many homes.

Government officials knew that New Orleans's levees couldn't withstand a severe hurricane.

Levees Fail

Much of New Orleans is below sea level. It uses a system of walls called levees to prevent nearby lakes from flooding the city. But Katrina's heavy rain and storm surge were too much for the levees. Several levees broke. Floodwater flowed over the tops of several others. Billions of gallons of water rushed into the city.

After Hurricane Katrina, more than 60,000 people had to be rescued from rooftops and flooded homes.

Citywide Crisis

More than 80 percent of New Orleans was underwater by August 30. People were stranded for days. Helicopters rescued people from rooftops. Food and clean water were limited.

Recovery Efforts

National Guard troops began to hand out food and water on September 2. They evacuated survivors and began rebuilding the city's levees. The last of Katrina's floodwaters were pumped out of New Orleans on October 11.

Rebuilding the Levees

The US Army Corps of Engineers rebuilt the city's levees after Hurricane Katrina. The rebuilt levees held strong when Hurricane Ida hit Louisiana as a Category 4 storm in 2021.

Many people in New Orleans no longer had homes to return to. Some left the city permanently.

Finding Victims

As the floodwaters lowered, rescue teams found victims. Many had drowned. Others died from health problems. Some died from falls or accidents as they tried to escape the storm.

Thousands of rescuers helped in the aftermath of Hurricane Katrina.

Several memorials honor the victims of Hurricane Katrina.

A Deadly Storm

Hurricane Katrina killed nearly 1,400 people. Millions more lost their homes. The storm caused $161 billion in damage. It was one of the most damaging hurricanes that ever hit the United States.

The tropical depression that became Hurricane Sandy started as a wave off the west coast of Africa.

Moving through the Caribbean

Hurricane Sandy began as a tropical depression in the Caribbean Sea in October 2012. By October 24, Sandy was a Category 1 hurricane. The hurricane passed over Jamaica, the Dominican Republic, and Haiti. It triggered deadly mudslides that killed dozens in Haiti.

Storm Systems Combine

Hurricane Sandy moved north. Near the mid-Atlantic, Sandy ran into another storm. The two storm systems combined. A third storm from the north kept the hurricane close to the US East Coast.

Hurricane Sandy's Path

Hurricane Sandy affected people in Jamaica, Cuba, Haiti, the Dominican Republic, The Bahamas, and the United States.

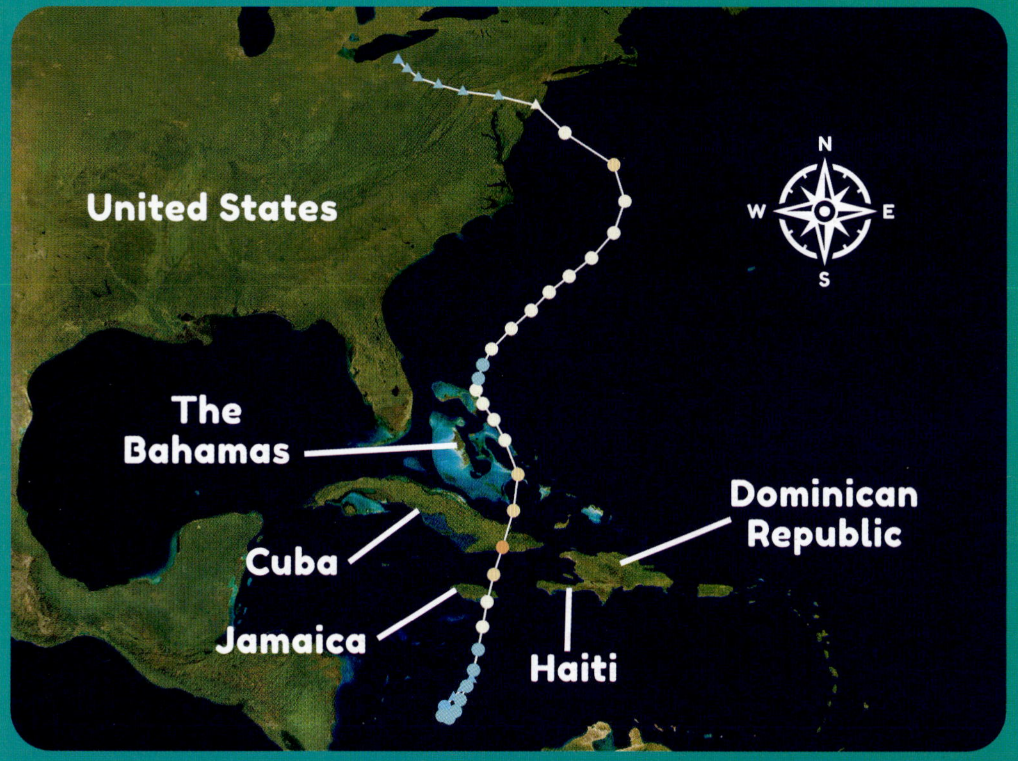

Landfall in New Jersey

Hurricane Sandy made landfall in New Jersey on October 29. It had sustained winds of 80 miles per hour (129 kmh). Atlantic City's famous boardwalk was damaged. In Hoboken, flooding trapped about 20,000 residents in their homes.

FUN FACT!

Hurricane Sandy's winds stretched as wide as 820 miles (1,320 km) at its peak.

Hurricane Sandy killed 12 people in New Jersey.

Hurricane Sandy killed 48 people in New York.

Flooding in New York City

The hurricane hit New York just hours later. Sandy's storm surge rose to almost 14 feet (4 m) in New York City. Rivers overflowed. Some of the city flooded. City subways and tunnels were underwater. Millions of people lost power.

Hospitals Evacuate

Flooding caused several New York City hospitals to lose power. They had to evacuate their patients.

President Barack Obama created a special task force to help rebuild after the damage caused by Hurricane Sandy.

A Destructive Hurricane

Hurricane Sandy killed 147 people. It caused more than $70 billion in damage. It was not the strongest hurricane to hit the United States. But it was one of the most destructive. Experts believe Sandy was so destructive because many cities were unprepared for it.

Unpreparred

New York's and New Jersey's floodwalls were out of date. Many people did not believe a storm like Sandy could hit so far north. This meant they were unprepared.

Rain and Snow

Hurricane Sandy brought record rainfall to the Northeast and mid-Atlantic. Some nearby states also received about three feet (1 m) of snow.

After Hurricane Sandy, New York City spent more than $1 billion building better floodwalls.

The Typhoon Forms

Super Typhoon Haiyan started as a tropical depression in November 2013. It moved west across the warm Pacific Ocean. It grew stronger and larger. It developed sustained winds of more than 150 miles per hour (240 kmh).

In the Philippines, Super Typhoon Haiyan is known as Yolanda.

The Philippines is located in the Pacific Ring of Fire. This area often gets natural disasters.

A Super Typhoon

Haiyan was classified as a super typhoon. A super typhoon is like a Category 4 or Category 5 hurricane. It was more than 500 miles (800 km) wide.

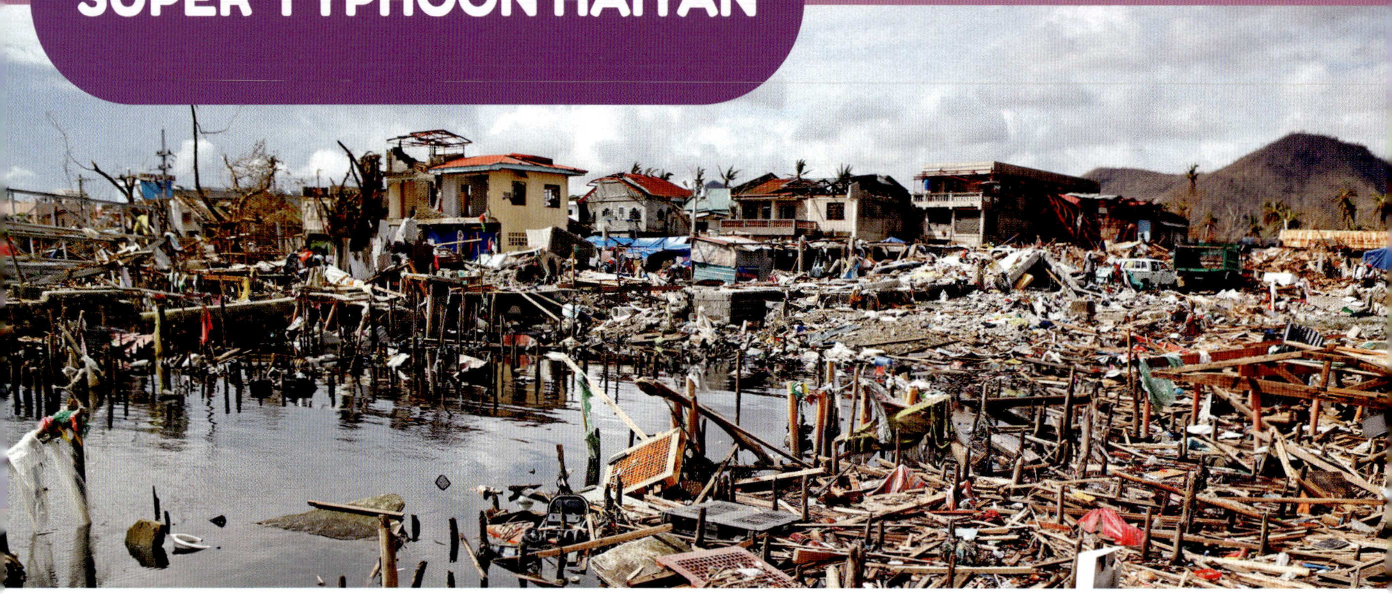

Super Typhoon Haiyan first made landfall on the island of Samar.

Landfall in the Philippines

Haiyan made landfall in the Philippines on November 8. It had wind speeds of 195 miles per hour (314 kmh). These are some of the strongest winds ever recorded at landfall. The typhoon caused power outages across the islands. The storm was so powerful that even government storm shelters were destroyed.

FUN FACT!

The Philippines averages about 20 typhoons every year.

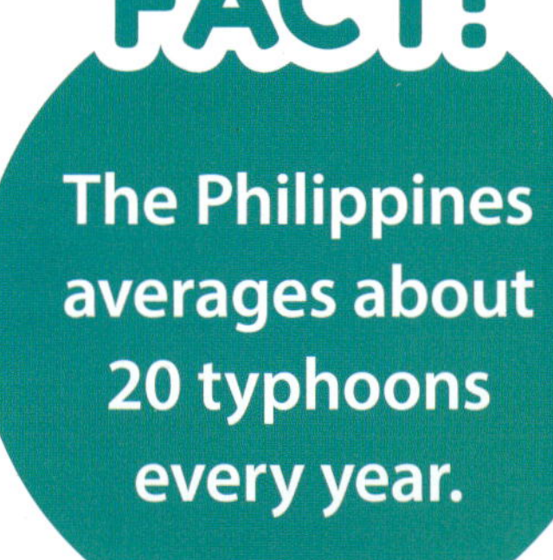

Destruction across the Islands

The Philippine islands of Samar and Leyte were very hard hit. Haiyan's storm surge rose 24 feet (7 m) high. It tossed boats. It destroyed buildings. It swept people and debris into the ocean.

Super Typhoon Haiyan's Path

Super Typhoon Haiyan affected more than 16 million people.

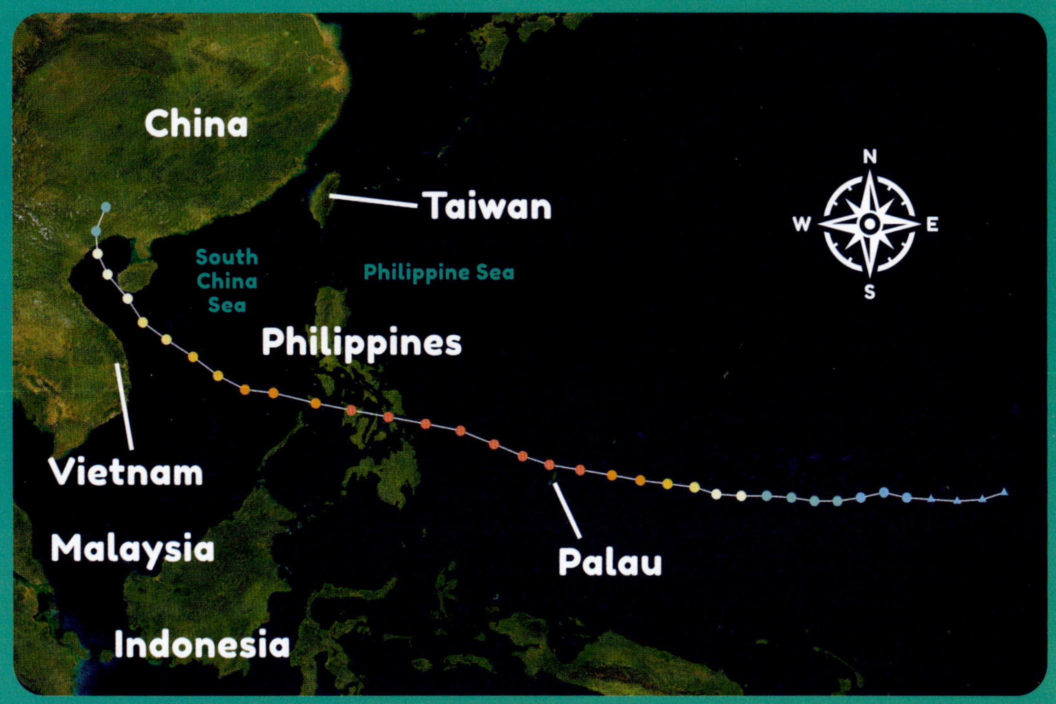

Rescue Efforts

Rescuers had trouble reaching some of the areas hit hardest by Haiyan. Aid workers had to clear roads before they could bring supplies. Several countries and relief organizations helped.

More than 50 countries provided aid after Super Typhoon Haiyan.

More than 600,000 people were evacuated in Vietnam.

Weakening Storm

Haiyan moved northwest. It traveled over the South China Sea. It made landfall in Vietnam on November 10 as a tropical storm.

SUPER TYPHOON HAIYAN

Super Typhoon Haiyan left more than four million people without homes.

Deadly Impact

Super Typhoon Haiyan destroyed approximately 550,000 homes. Another 580,000 homes were severely damaged. More than 6,000 people died. Thousands more were injured. It was one of the strongest typhoons ever recorded.

Higher Sea Levels

Rising sea levels made Super Typhoon Haiyan more destructive. Sea levels near the Philippines have risen nearly eight inches (20 cm) since the 1970s. This added to the typhoon's storm surge and coastal flooding.

An Unusual Storm

Super Typhoon Haiyan was an unusual storm. It was extremely powerful. It also formed close to the equator. This is rare for tropical cyclones.

Relief organizations around the world helped people in the Philippines build safer homes.

GLOSSARY

buoy
A floating object anchored in a body of water.

classify
To sort things into groups according to their characteristics.

condense
To change from a gas to a liquid.

debris
The remains of something broken or destroyed.

equator
An imaginary line that divides Earth into northern and southern halves.

erosion
A process in which soil and rock are worn away and moved by wind or water.

evacuate
To remove someone from a dangerous place to a safer place.

forecast
To predict a future event.

glacier
A large, slow-moving mass of ice.

humid
Having a high amount of water vapor in the air.

meteorologist
A scientist who studies weather.

radar
A system that uses radio waves to detect objects.

satellite
A human-made device that orbits Earth or another body in space.

sustained winds
The average wind speed over two minutes.

uninhabitable
Unable to be lived in.

TO LEARN MORE

More Books to Read

Buckey, A. W. *Weather*. Abdo, 2024.

Challoner, Jack. *Hurricane and Tornado*. DK, 2021.

Watts, Claire. *Eyewitness Natural Disasters*. DK, 2022.

Online Resources

To learn more about hurricanes, please visit **abdobooklinks.com** or scan this QR code. These links are routinely monitored and updated to provide the most current information available.

INDEX

PHOTO CREDITS

Cover Photos: Shutterstock Images, front (boy, palm tree); Roberto Machado Noa/Moment/Getty Images, front (hurricane); Mark Winfrey/Shutterstock Images, back

Interior Photos: Shutterstock Images, 1, 3, 9, 11, 12, 15, 23, 24, 27 (compass), 28, 29 (top), 29 (bottom), 33 (top), 43 (bottom), 45 (satellite), 46, 49, 53 (building), 53 (background), 56, 59 (top), 65, 66, 68, 76, 78 (top), 80, 81, 82, 84, 90, 91 (bottom), 102, 109 (top), 112, 115 (bottom), 117 (top), 119, 120, 122, 125 (top); Vladislav Gurfinkel/Shutterstock Images, 4; Win McNamee/Getty Images News/Getty Images, 5; iStockphoto, 6, 8, 10; Dorling Kindersley/Dorling Kindersley RF/Getty Images, 7; NASA, 13, 16, 36, 118; Antti Lipponen/Flickr, 14; Roberto Machado Noa/Moment/Getty Images, 17; NOAA, 18, 22, 35, 42, 44, 45 (except satellite), 72, 88; Larry Rains/Shutterstock Images, 19; Sean Rayford/Getty Images News/Getty Images, 20; Mark Wallheiser/Getty Images News/Getty Images, 21; Library Book Collection/Alamy, 25; Red Line Editorial/NOAA, 26, 103; Red Line Editorial, 27 (map), 94; Joe Raedle/Getty Images News/Getty Images, 30, 50, 52, 63, 71, 77, 89, 93; NASA Goddard Space Flight Center, 31, 34, 37; Andrei Armiagov/Shutterstock Images, 32; Viacheslav Lopatin/Shutterstock Images, 33 (bottom); Air Force Reserve/AP Images, 38; Gemunu Amarasinghe/AP Images, 39; Marie D. De Jesus/Houston Chronicle/Hearst Newspapers/Getty Images, 40; Science History Images/Alamy, 41; rulenumberone2/Flickr, 43 (top); Orlando Sierra/AFP/Getty Images, 47; Felix Mizioznikov/Shutterstock Images, 48; Scott Olson/Getty Images News/Getty Images, 51, 96; Warren Faidley/The Image Bank/Getty Images, 54; Justin Hobson/Shutterstock Images, 55; Donna Bollenbach/Shutterstock Images, 57; Rudmer Zwerver/Shutterstock Images, 58; Mark Winfrey/Shutterstock Images, 59 (bottom); Tom MacKenzie/USFWS, 60; USFWS, 61; Sandy Huffaker/Getty Images News/Getty Images, 62; Leonard Zhukovsky/Shutterstock Images, 64, 116; Jillian Cain Photography/Shutterstock Images, 67; Alleymaher/Dreamstime, 69; Paul Hennessy/SOPA Images/LightRocket/Getty Images, 70; Richard Unten/Flickr, 73; Susan Watts/New York Daily News/Getty Images, 74; Darwin Brandis/iStockphoto, 75; Marko Georgiev/Getty Images News/Getty Images, 78 (bottom); Spencer Platt/Getty Images News/Getty Images, 79; Cpl. Samuel Guerra/US Marine Corps/DVIDS, 83; Matthew Jolley/Shutterstock Images, 85; Arterra/Universal Images Group/Getty Images, 86; Tom Wang/Shutterstock Images, 87; Chip Somodevilla/Getty Images News/Getty Images, 91 (top); Jason Edwards/The Image Bank/Getty Images, 92; Scott Pena/Flickr, 95; Brandon Bell/Getty Images News/Getty Images, 97; NOAA/NASA, 98; Todd Maisel/New York Daily News/Getty Images, 99, 115 (top); switas/E+/Getty Images, 100; Ricardo Arduengo/AFP/Getty Images, 101; Mario Tama/Getty Images News/Getty Images, 104, 108; Dave Martin/AP Images, 105; Chris Hondros/Getty Images News/Getty Images, 106; Jerry Grayson/Helifilms Australia PTY Ltd/Getty Images News/Getty Images, 107; Robyn Beck/AFP/Getty Images, 109 (bottom); Chris Graythen/Getty Images News/Getty Images, 110; Dimitry Bobroff/Alamy, 111; Wikimedia Commons, 113, 121; Glynnis Jones/Shutterstock Images, 114; Ed Jones/AFP/Getty Images, 117 (bottom); STR/AFP/Getty Images, 123; Jan Hetfleisch/Getty Images News/Getty Images, 124; Aaron Favila/AP Images, 125 (bottom)